Cyclists Please Dismount

# Cyclists Please Dismount

And other photographs from Kodak Limited's "Humour '70" Exhibition

Foreword by Spike Milligan

ANGUS AND ROBERTSON

*First published in 1971 by*
ANGUS AND ROBERTSON (PUBLISHERS) PTY LTD

2 Fisher Street, London, WC1
221 George Street, Sydney
107 Elizabeth Street, Melbourne
89 Anson Road, Singapore

ISBN 0 207 95440 2

The publishers gratefully acknowledge the help and
co-operation of Kodak Ltd, London, in the preparation
of this book, and of the photographers whose work is
included. The photographs published here were selected
from the "Humour '70" Exhibition at Kodak House,
Kingsway, London, in 1970. The judges of the exhibition
were William Davis, Frank Muir and Spike Milligan.

The photograph on the front cover is by E. Davis.

Made and printed in Great Britain by
Ebenezer Baylis & Son Ltd, Leicester and London

# Foreword

No one dreamed what lay ahead that distant day in 1802 (yes, 1802) when Thomas Wedgwood told the world the results of his experiments in recording photographic images. Photography remained in a state of obscurity until the 1830s, when Fox Talbot and Daguerre started to take pictures in earnest. The camera soon became the poor man's portrait-painter; people who could never have afforded an oil painting could now indulge in photographs of their families to put on top of the withdrawing room piano, where they were lovingly dusted once a week. The difficulties of handling the temperamental collodion wet plate and the sensitive daguerreotype kept the profession of photography for a long time in the hands of those who studied its chemistry; but then, in the 1880s – BOOM! like Ford's Tin Lizzie, George Eastman produced the first cheap camera using roll film.

At first the price of Eastman's box Kodak only allowed middle-class access, but as the new century grew older cameras became cheaper, and by the thirties the "Brownie" had become available to the most humble of families. But the real boom came after World War Two. As the new affluent society flexed its limbs, nearly every family acquired a camera; no tourist would consider himself on holiday without one slung round his neck. Today, with the instant camera, you can shoot anything anywhere and have the results in your hands within seconds. The camera has changed man's life and improved his memory. I myself have photographs that I will always cherish – my kids in the bath, playing in the garden, holding the puppy, asleep in bed, etc., etc.

Now this book. Life is a series of serio-comic adventures, if you are observant (and if you want to be a photographer you've got to be). Not a day passes but what we see some comic incident. Often the actual moment of the comedy is over, like a wisecrack, in a flash. (Or in a flash camera! Joke.)

In this book we see a wide range of very different subjects, but they all have one thing in common: they are very funny. I sat on the panel of judges to choose the pictures for Kodak's Humour '70 Exhibition. I had never sat in judgment of my fellow men's efforts before, and I was surprised at the number of entries. Ploughing through we came across large numbers of photographers who thought that the mere inclusion of a chamber pot was sufficient to make a picture funny. There were others on cruises who thought that for three people to stand on deck with funny paper hats on was hilarious. And there was the usual quota of photographs with Ladies and Gents lavatory signs in the background, and so on. The photographs you see here have been chosen from what the publishers considered the best in the exhibition. They are a selection from the original thousands of entries.

I have looked carefully through the contents and I must say that no matter at what page I opened the book, I always had a giggle or a belly laugh. The book is best viewed with someone else. It has to be shared, as all laughter should be. There are millions of books full of gloom, religion, murders, etc., etc., and far too few that give us that God-given sound of laughter.

This book has no gimmicks, no messages, just fun. What more can you ask of a book?

SPIKE MILLIGAN
*London, May 1971*

R. F. Tredwen

Roy Palmer

R. D. Geere

9

D. G. Stuart

A. Lagocki

M. C. Quinton

Roger Canessa, AFIAP

M. J. Crispin

H. H. Morris

G. Rywacki

B. Fretton

Mike Holmes

Jane Miller

A. N. A. Argent

19

Fox Photos
Fox Photos

Serge de Sazo

Ron Dumont, *Daily Express*

F. RUDE
1784 - 1855

Richard H. Bomback, FRPS

23

Derek L. Harding

TO THE MEMORY

OF

THOMAS LEWIS OF SAINT PIERRE ESQ.ʳ

WHO DIED

JUNE 17ᵀᴴ A:D:1796, AGED 45.

THIS MONUMENT WAS ERECTED

BY

HIS GRATEFUL WIDOW.

Eric Ambrose

Dr W. R. M. Thompson

P. W. Lyons

Dorothy Cloynes
T. L. Ralph

C. J. Benike-Oakeley

H. Farrer

Cyril Bernard

G. Crosby, FRPS

Dorothy Cloynes
Neville Ian Ash

Mrs I. L. Cochrane

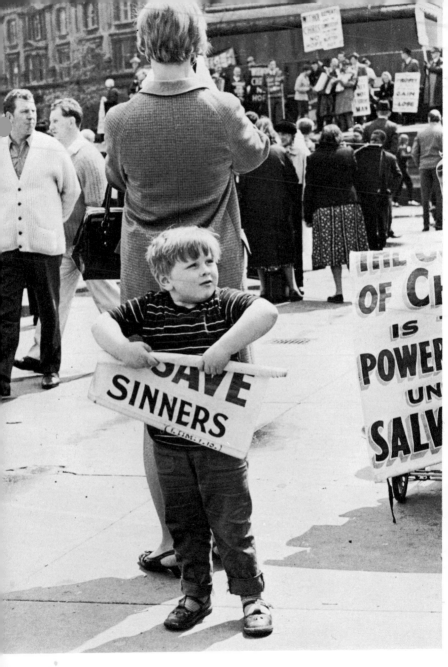

Derek Shuff

A. N. Hooker

Ron Chapman, AFIAP

Peter Hunter, ARPS, AFIAP

PLEASE
DO NOT PARK
ON GRASS

NO BALLS·NO DOGS·
NO RADIOS·NO FLIPPERS
NO BOMBING
It is regretted that we have to
make these prohibitions, but they
are necessary for the public
safety and convenience.

Larry Beale
Louis and Daphne Peek

Neville Ian Ash

Ron Chapman, AFIAP

40

Fox Photos

LUXURY
GUEST HOUSE

WILL BE OPENED
AFTER RESTORATION

Developers:
CANFORD GARDEN ESTATE LTD.
Contractors:
J.W. WILES LTD.

Keith R. Hughes

Fox Photos

BANK OF SCOTLAND

PUBLIC RELATIONS

DEPARTMENT

EDINBURGH
DESTITUTE SICK
SOCIETY

Neville Ian Ash

D. Adams

V. Lakey

V. Lakey

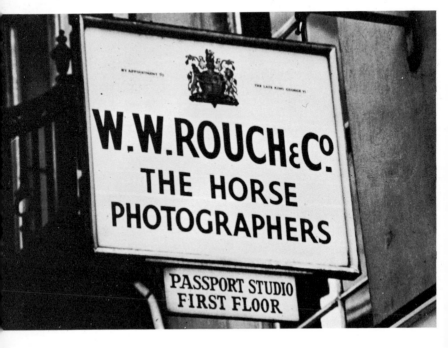

B. Fretton

F. J. Turner

Fox Photos

Larry Beale

George Jenkinson, ARPS

Colin Earl and Ken Lauder

Colin Earl and Ken Lauder

J. H. White

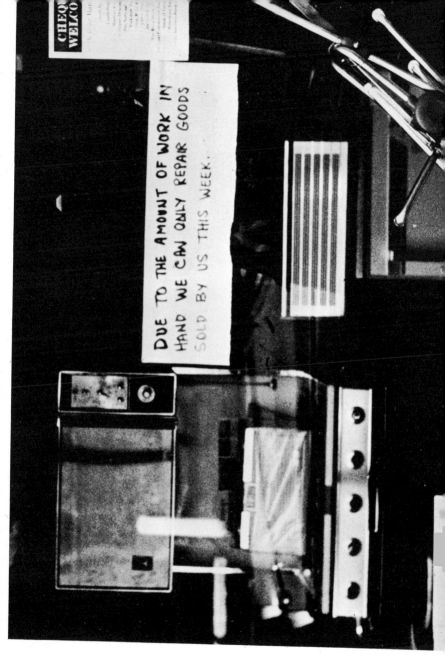

Mrs J. Counsell

Sheelah Latham, ARPS, AFIAP

George Jenkinson, ARPS

Ron Chapman, AFIAP

P. Callaghan

D. M. Barnard

J. H. White

TO ENTER THIS FIELD IS DANGEROUS —
UNLESS YOU CAN RUN 100 YDS IN 9 SECS.
MY BOAR DOES IT IN 10 SECS. FLAT

OFFI...

F. A. Mays

George Jenkinson, ARPS

P. Dowles

Paul Hill